Copyright (c) 2024 by Joy Semien

All rights reserved. No part of this book may be reproduced, distributed, or transmitted in any form or by any means, including photocopying, recording, or other electronic or mechanical methods, without the prior written permission of the author, except in the case of brief quotations embodied in reviews and other specific non-commercial uses permitted by copyright law.

Publisher:
THGM Publishing

P.O. Box 562 Geismar, Louisiana 70734

https://www.thekapsdisasterhub.com/

THIS BOOK BELONGS TO:

_____

Tracing Dinos!

Tracing Dino Numbers

1 2 3 4

5 6 7 8

9 0

Tracing Dino Numbers

0	0	0	0
0	0	0	0
0	0	0	0
0	0	0	0
0	0	0	0

Tracing Dino Numbers

Tracing Dino Numbers

Tracing Dino Numbers

Tracing Dino Numbers

Tracing Dino Numbers

5	5	5	5
5	5	5	5
5	5	5	5
5	5	5	5
5	5	5	5

Tracing Dino Numbers

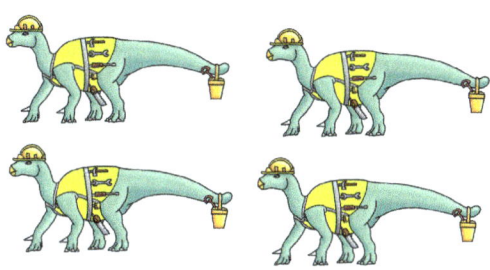

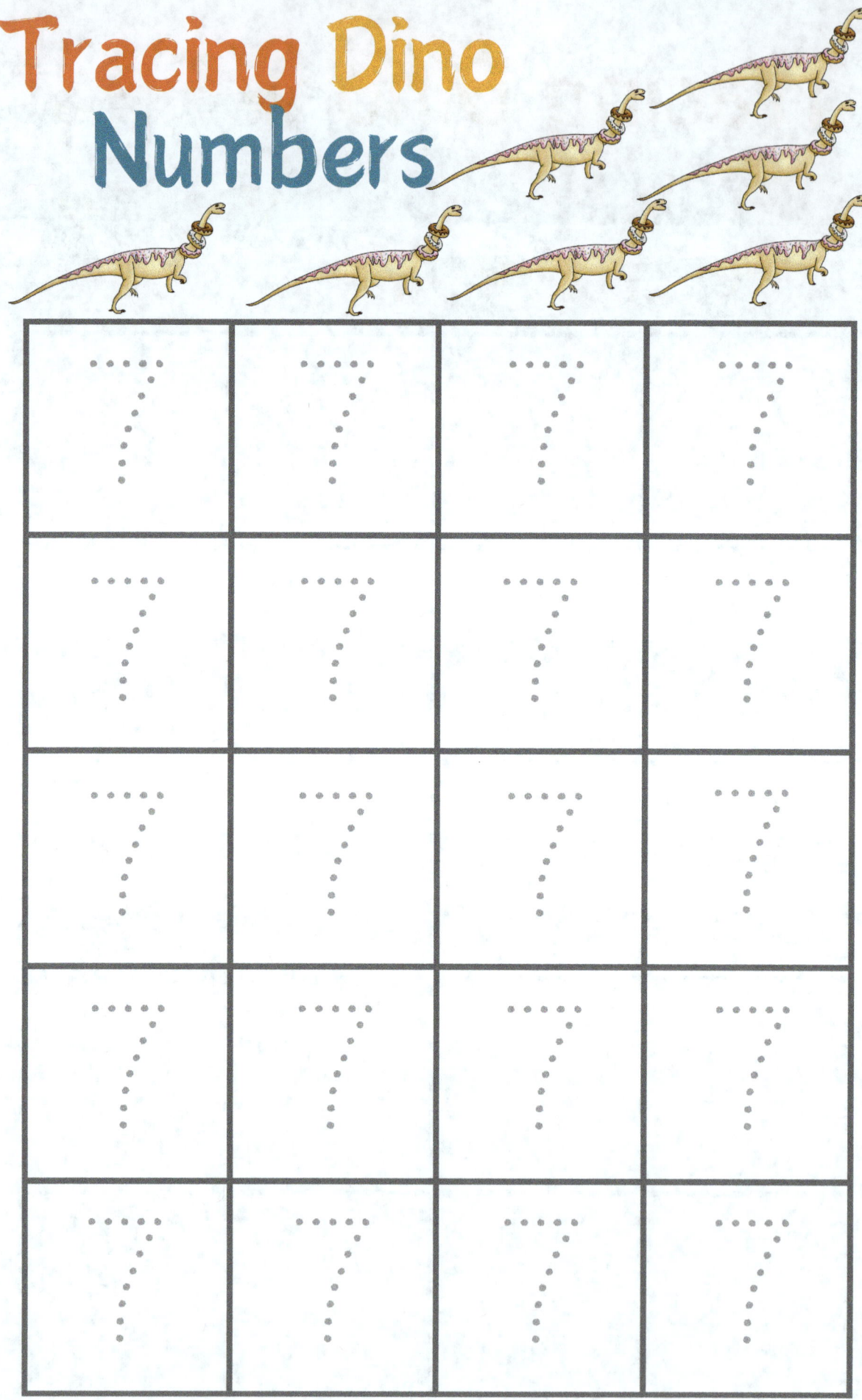

Tracing Dino Numbers

Tracing Dino Numbers

Tracing Dino Numbers

	1 1 1 1 1 1
	2 2 2 2 2
	3 3 3 3 3
	4 4 4 4 4
	5 5 5 5 5

Dino Fun

Let's Count Dinos!

Let's Count!

Count the Dinos and trace the number on the line.

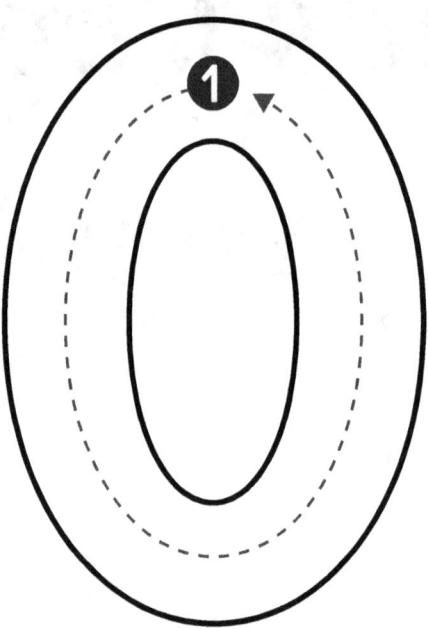

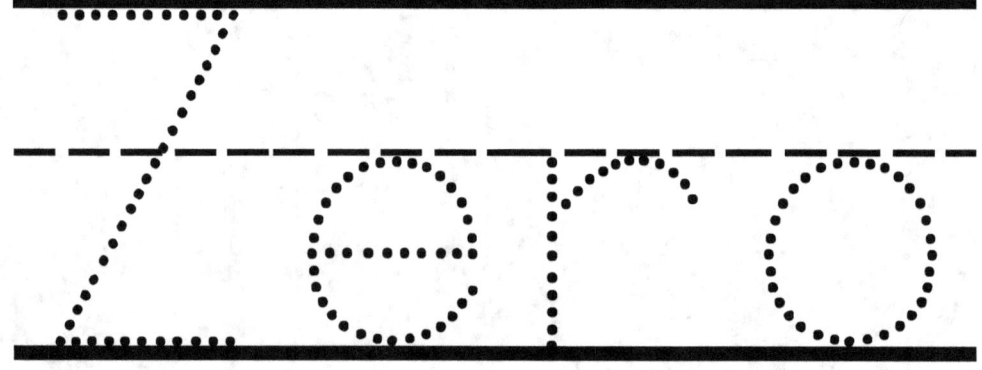

Let's Count!

Count the Dinos and trace the number on the line.

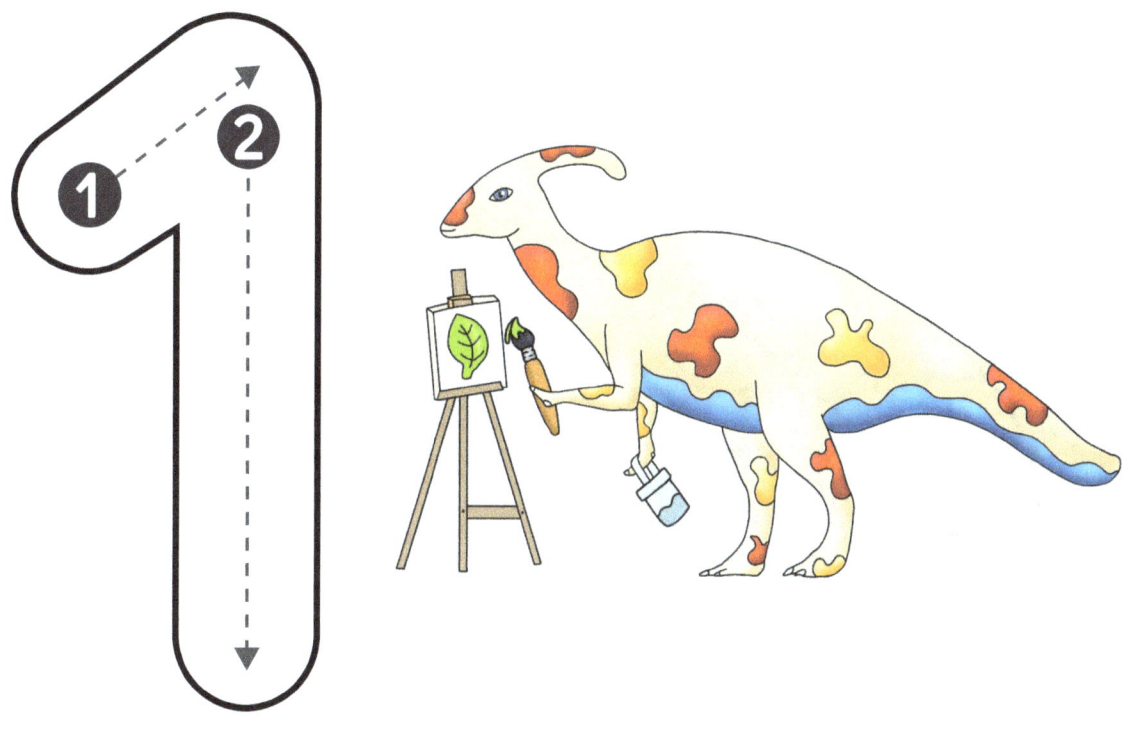

Let's Count!

Count the Dinos and trace the number on the line.

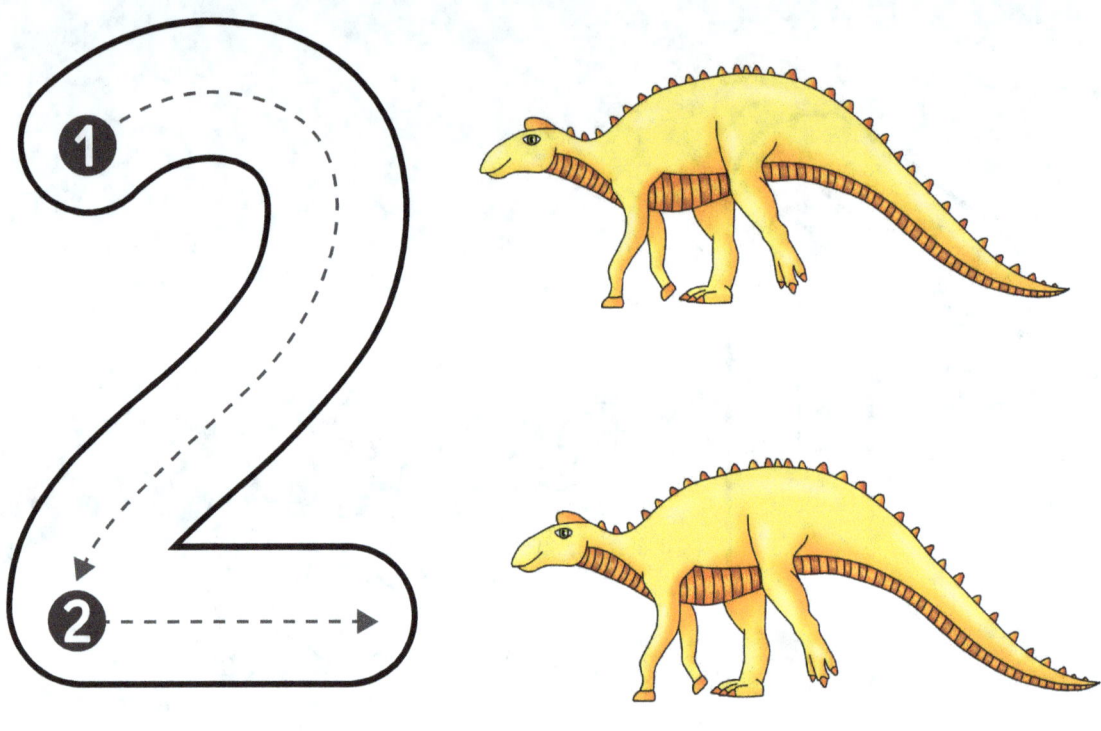

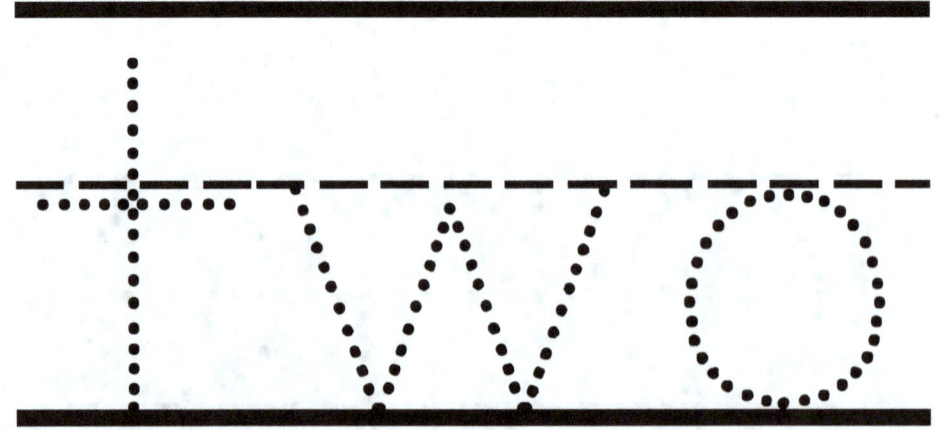

Let's Count!

Count the Dinos and trace the number on the line.

Let's Count!

Count the Dinos and trace the number on the line.

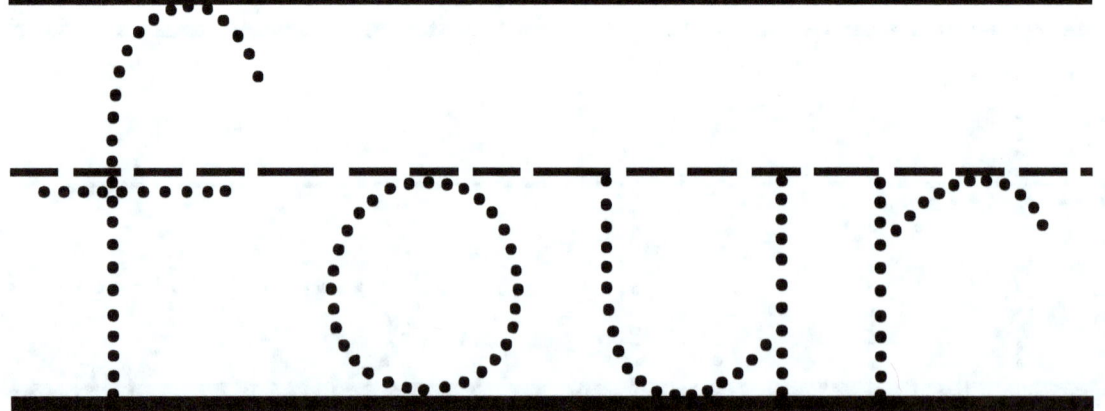

Let's Count!

Count the Dinos and trace the number on the line.

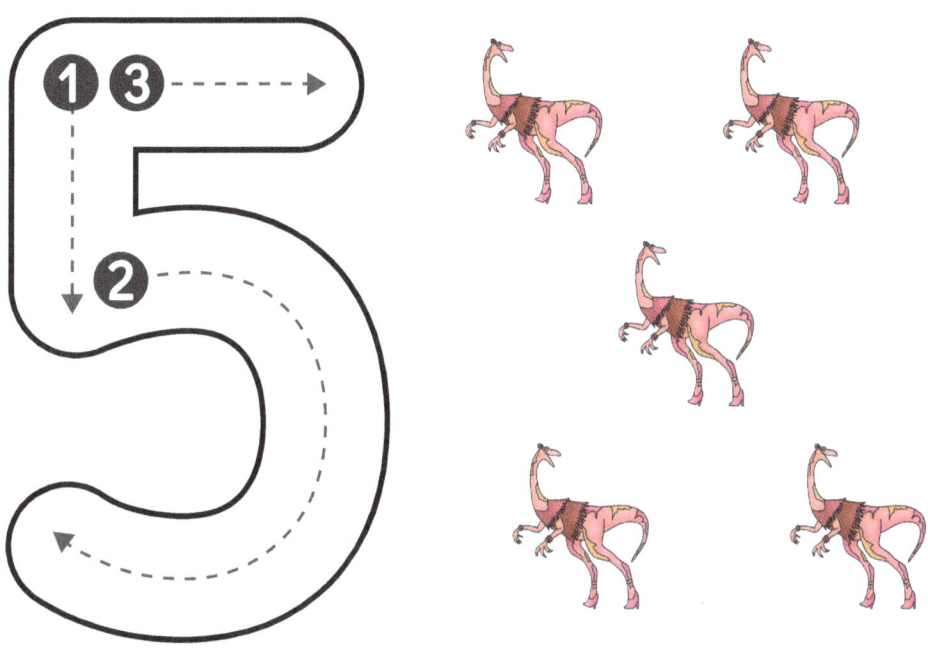

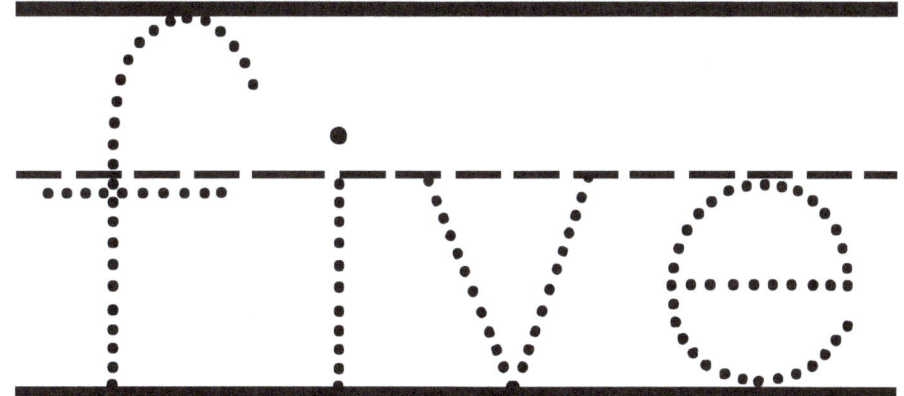

Let's Count!

Count the Dinos and trace the number on the line.

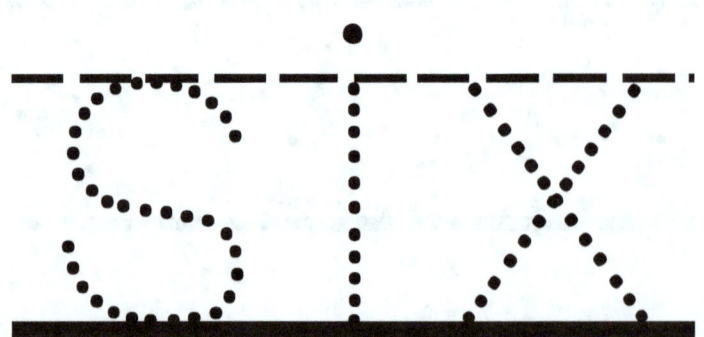

Let's Count!

Count the Dinos and trace the number on the line.

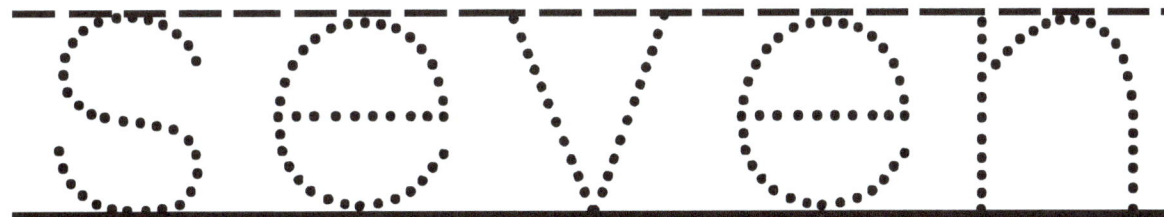

Let's Count!

Count the Dinos and trace the number on the line.

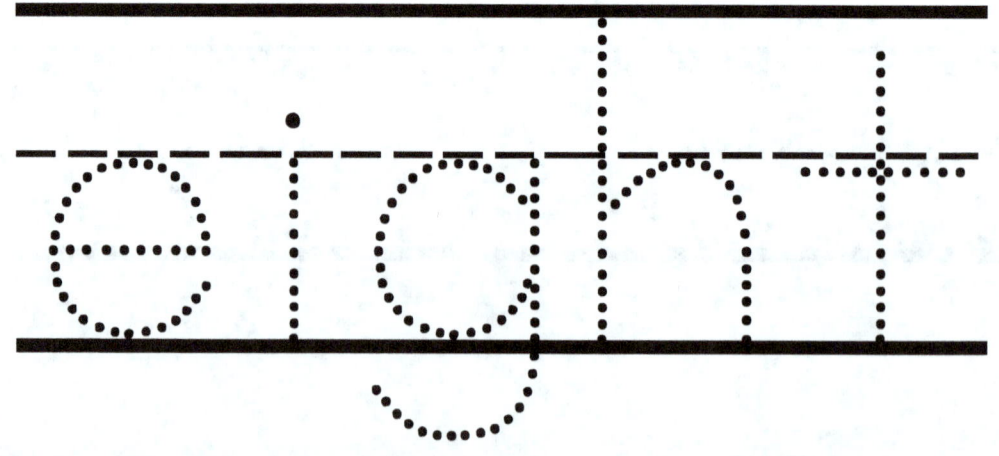

Let's Count!

Count the Dinos and trace the number on the line.

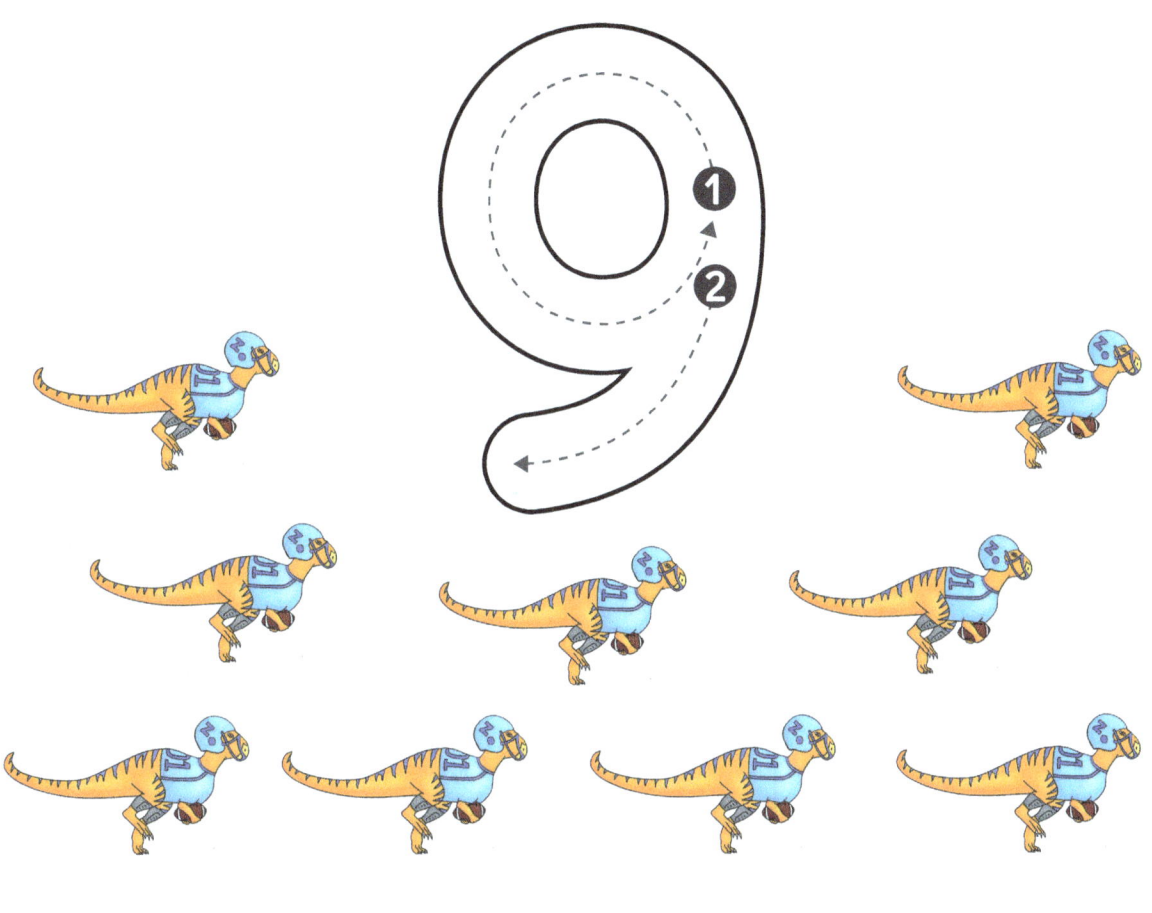

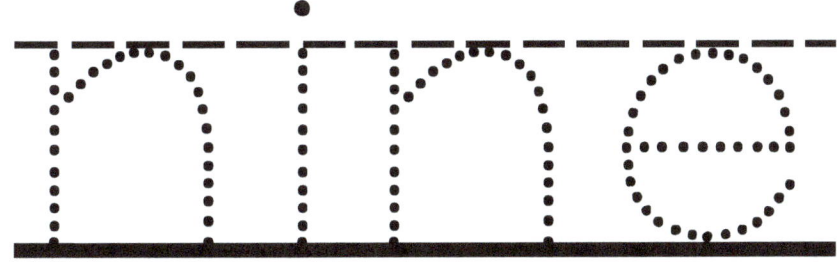

Let's Count!

Count the Dinos and trace the number on the line.

Dino Counting!

Dino Counting

Count the Dino and write the number in the box.

Dino Counting

Count the Dino and write the number in the box.

Dino Counting

Count the Dino and write the number in the box.

Dino Counting

Count the Dino and write the number in the box.

Dino Counting

Count the Dino and write the number in the box.

Dino Counting

Count the Dino and write the number in the box.

Dino Counting

Count the Dino and write the number in the box.

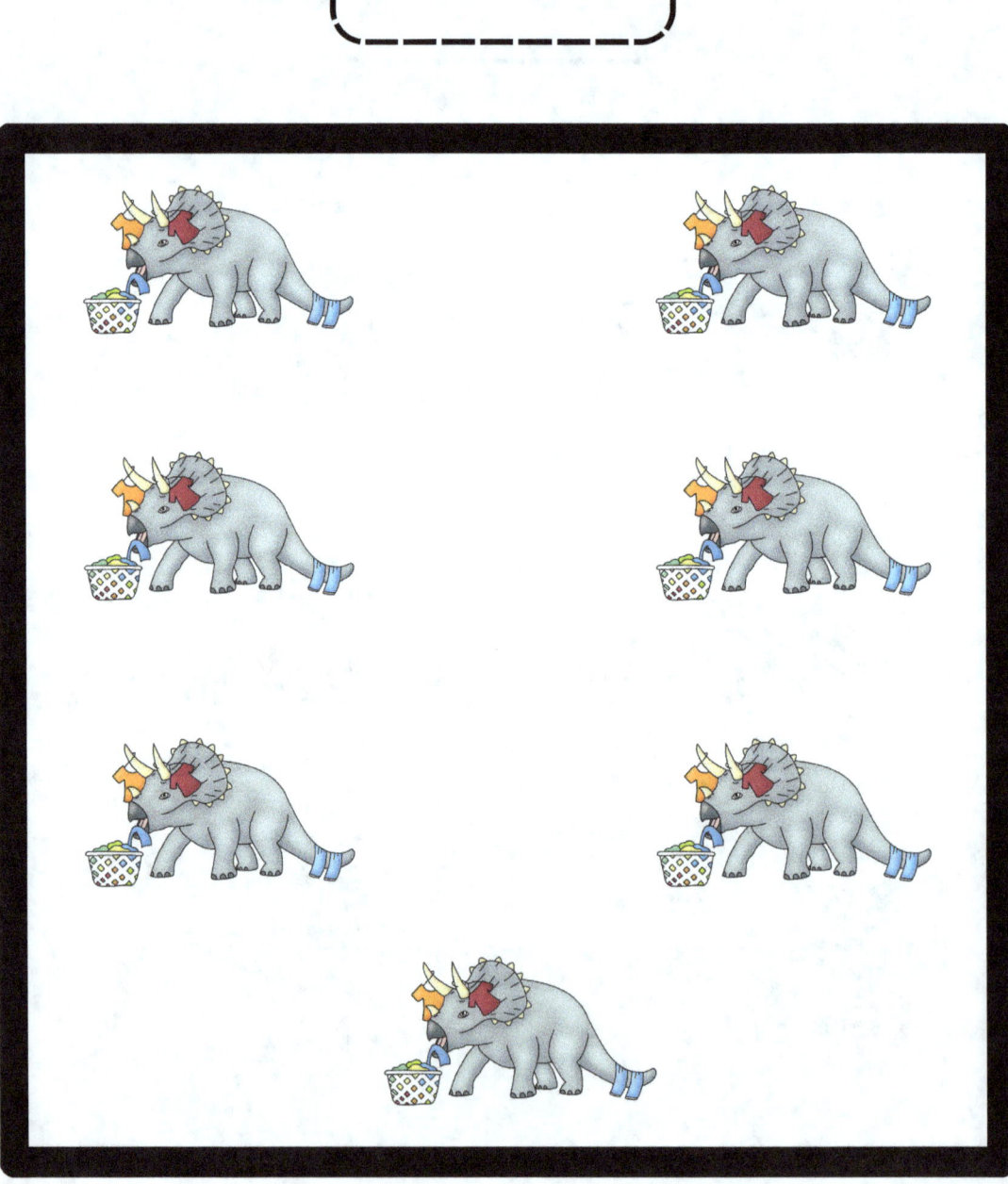

Dino Counting

Count the Dino and write the number in the box.

Dino Counting

Count the Dino and write the number in the box.

Dino Counting

Count the Dino and write the number in the box.

Dino Fun

Let's Count & Color Dinos!

Count And Color

Count the objects and write the number in the box.

Count And Color

Count the objects and write the number in the box.

Count And Color

Count the objects and write the number in the box.

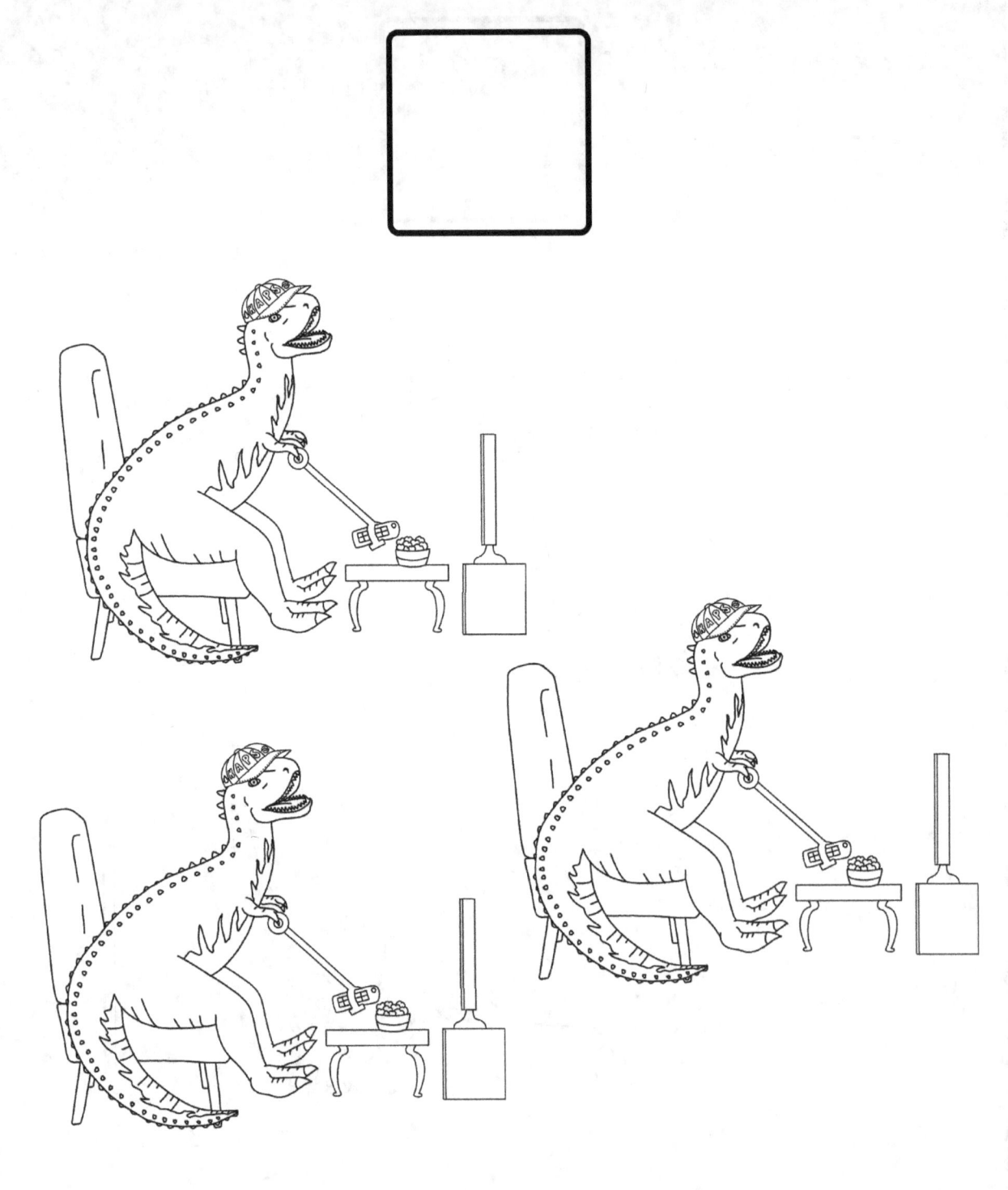

Count And Color

Count the objects and write the number in the box.

Count And Color

Count the objects and write the number in the box.

Count And Color

Count the objects and write the number in the box.

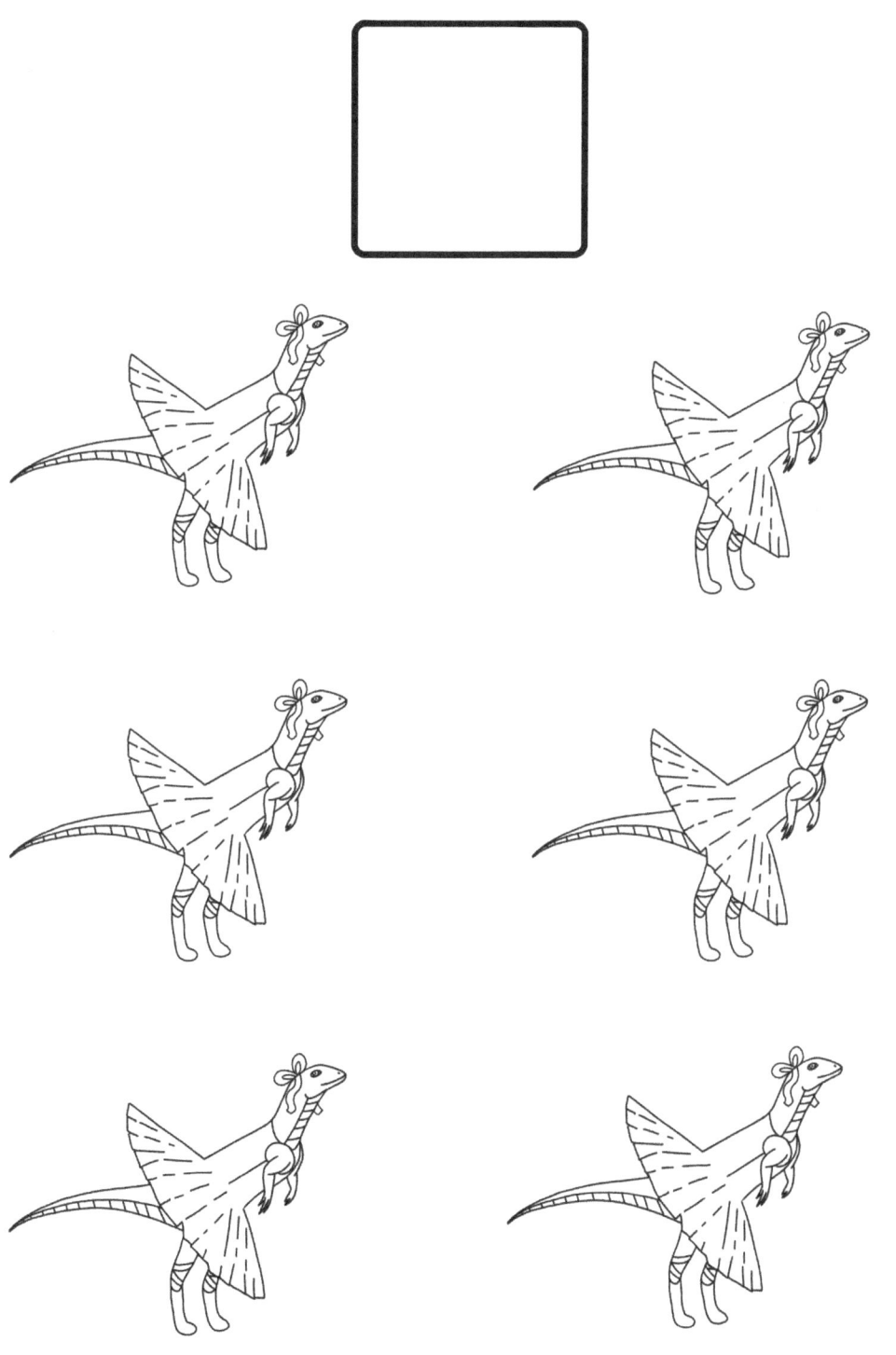

Count And Color

Count the objects and write the number in the box.

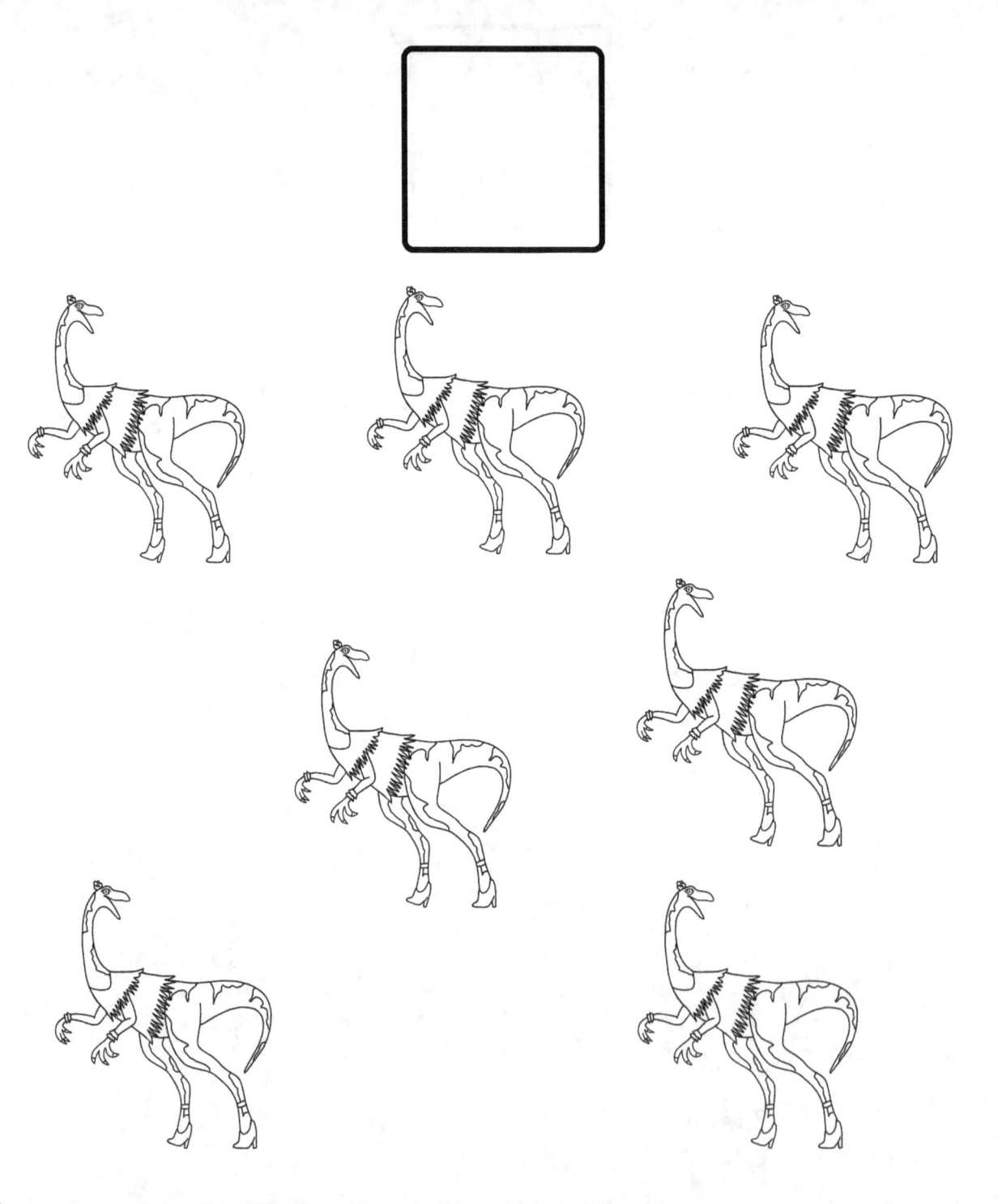

Count And Color

Count the objects and write the number in the box.

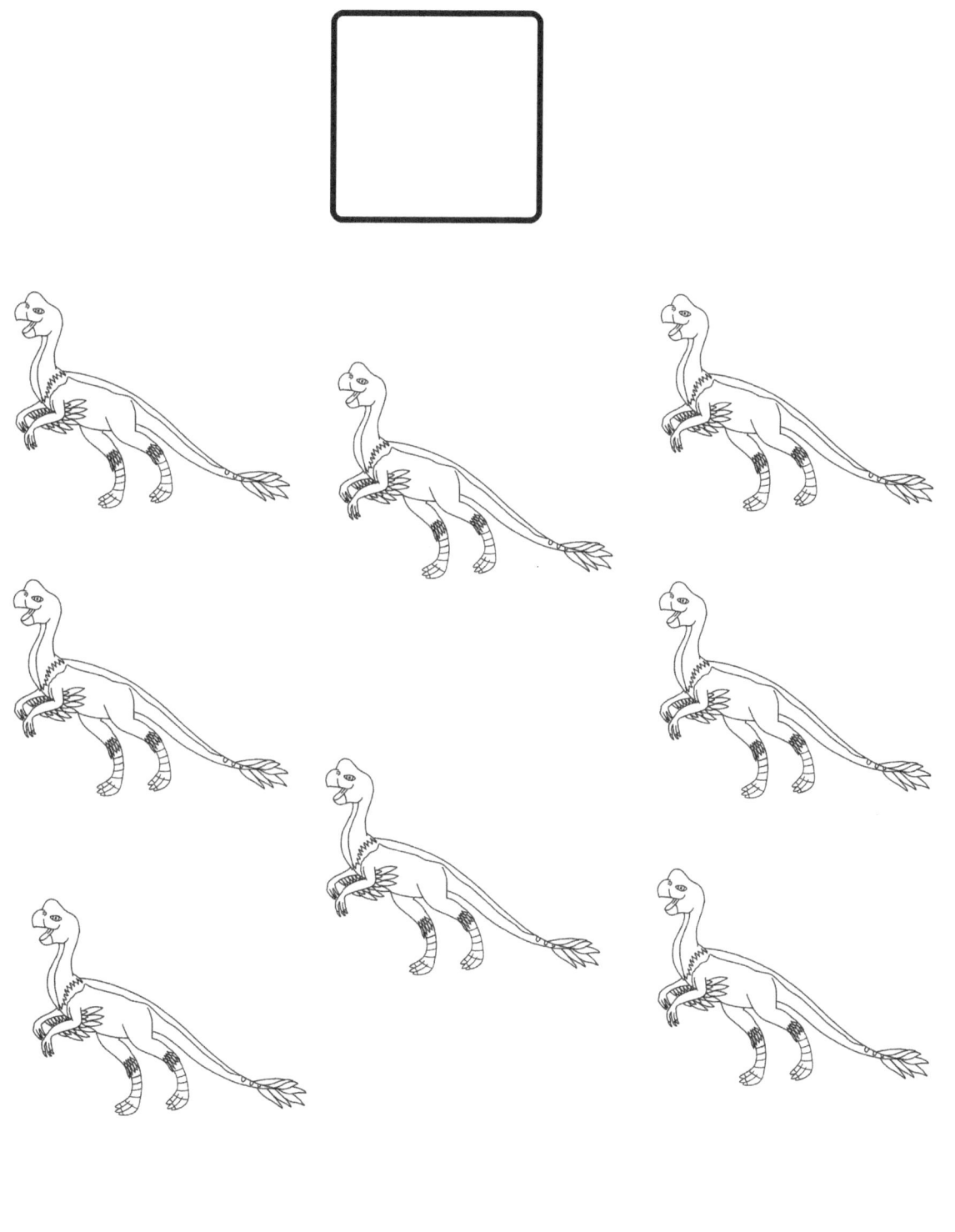

Count And Color

Count the objects and write the number in the box.

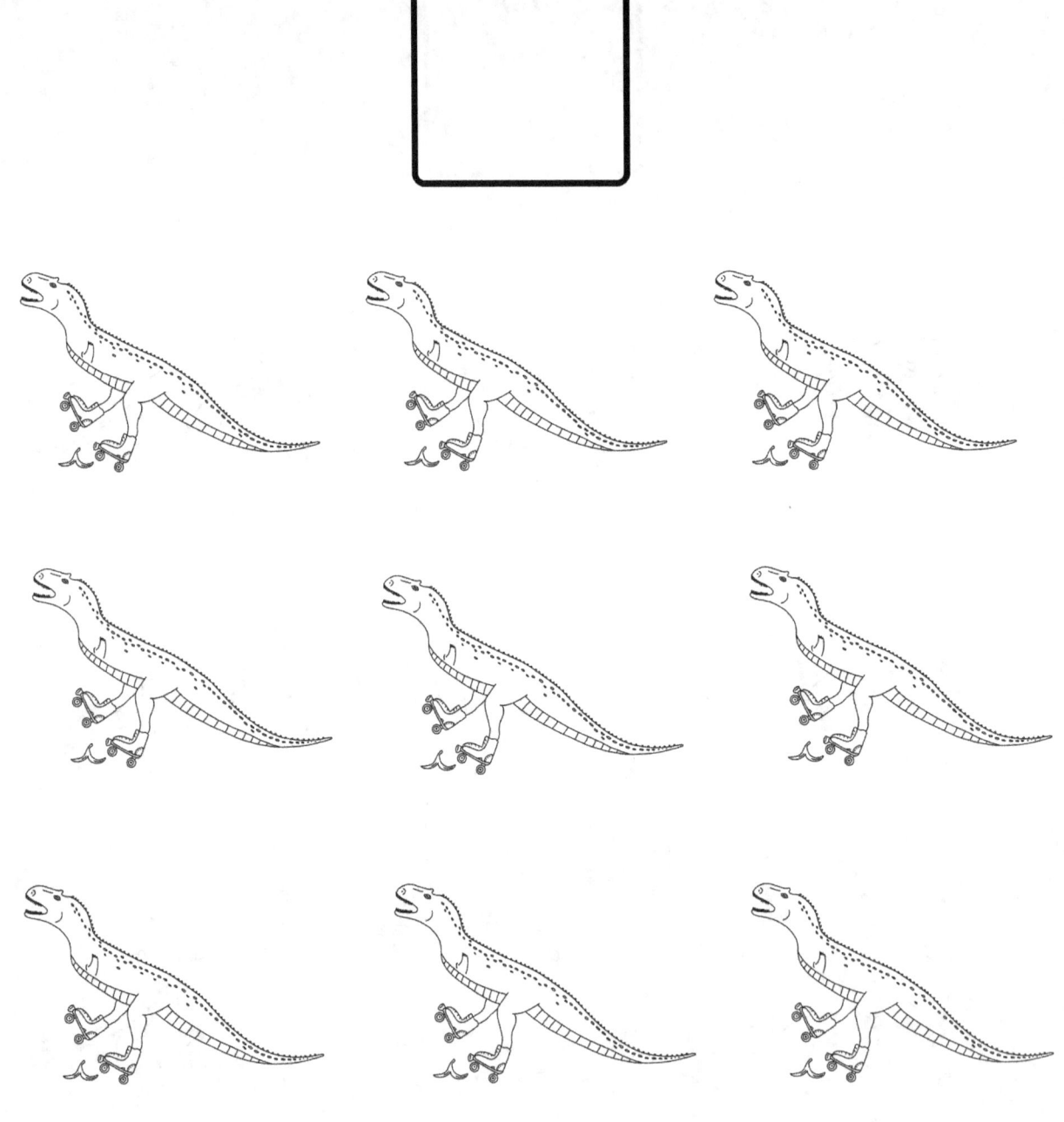

Count And Color

Count the objects and write the number in the box.

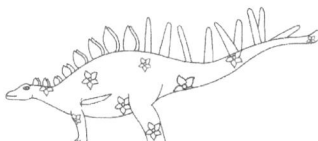  

Dino Fun

Let's Count, Color, & Write

Count, Color, & Write

Count the Dinos and write the numbers (1-8) in the boxes

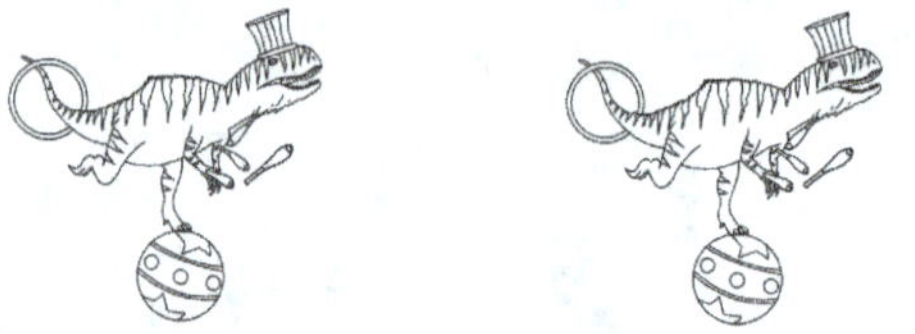

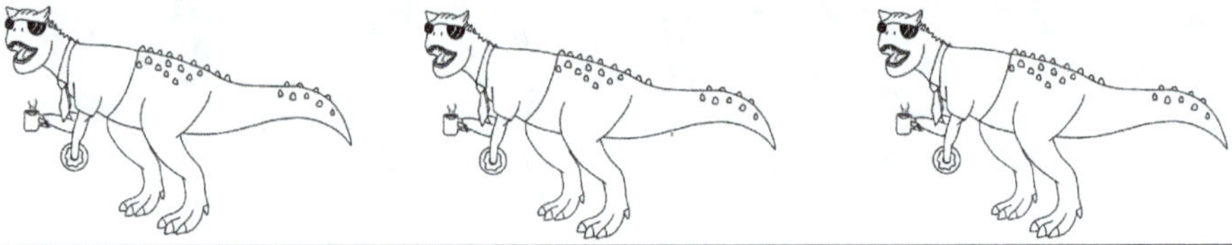

Count, Color, & Write

Count the Dinos and write the numbers (9-10) in the boxes

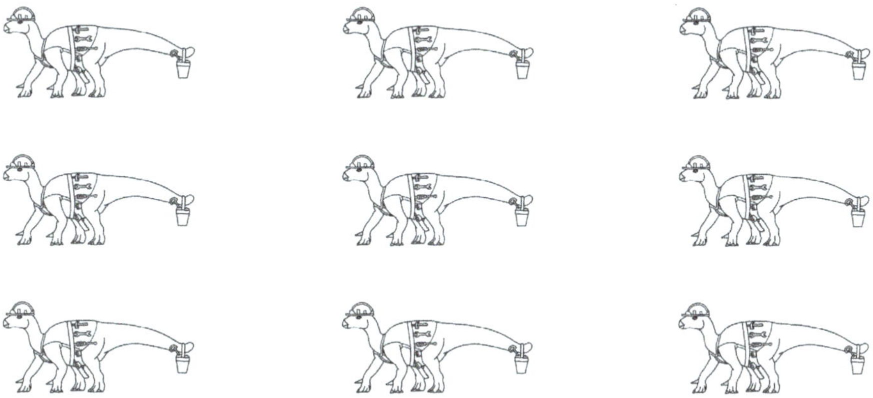

Count & Circle

Count the Dinos in each box and mark the correct number.

Count & Circle

Count the Dinos in each box and mark the correct number.

Count & Circle

Count the Dinos in each box and mark the correct number.

10 8 6 9

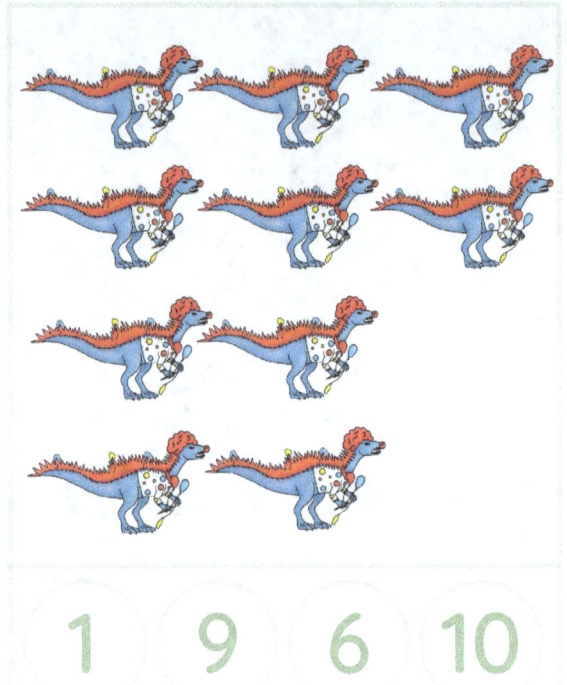

1 9 6 10

7 5 6 8

7 9 5 6

Count & Write

Count and write your answers in the chart below.

Dino Matching

Match the dinos.

Dino Matching

Match the dinos.

WHAT IS THE NUMBER?

What number is hidden under the picture?
Write the number in the box next to the picture.

1		3	4	5	6	7	8	9	10
11	12	13		15	16	17	18	19	20
21	22	23	24	25	26		28	29	30
31	32	33		35	36	37	38	39	40
41	42	43	44	45	46	47	48	49	

WHAT IS THE NUMBER?

What number is hidden under the picture?
Write the number in the box next to the picture.

1	2	3	4	5	6	7	8	9	10
	12	13	14	15	16	17	18		20
21	22		24	25	26	27	28	29	30
31	32	33	34	35	36	37	38	39	
41	42	43	44	45	46	47	48	49	50
51		53	54	55	56	57	58	59	60
61	62	63	64	65	66		68	69	70
71	72	73		75	76	77	78	79	80
81	82	83	84	85	86	87	88	89	90
	92	93	94	95	96	97	98	99	

Dino Fun

Dino Affirmations!

I AM
Kind

I AM Bold

I AM
Loved

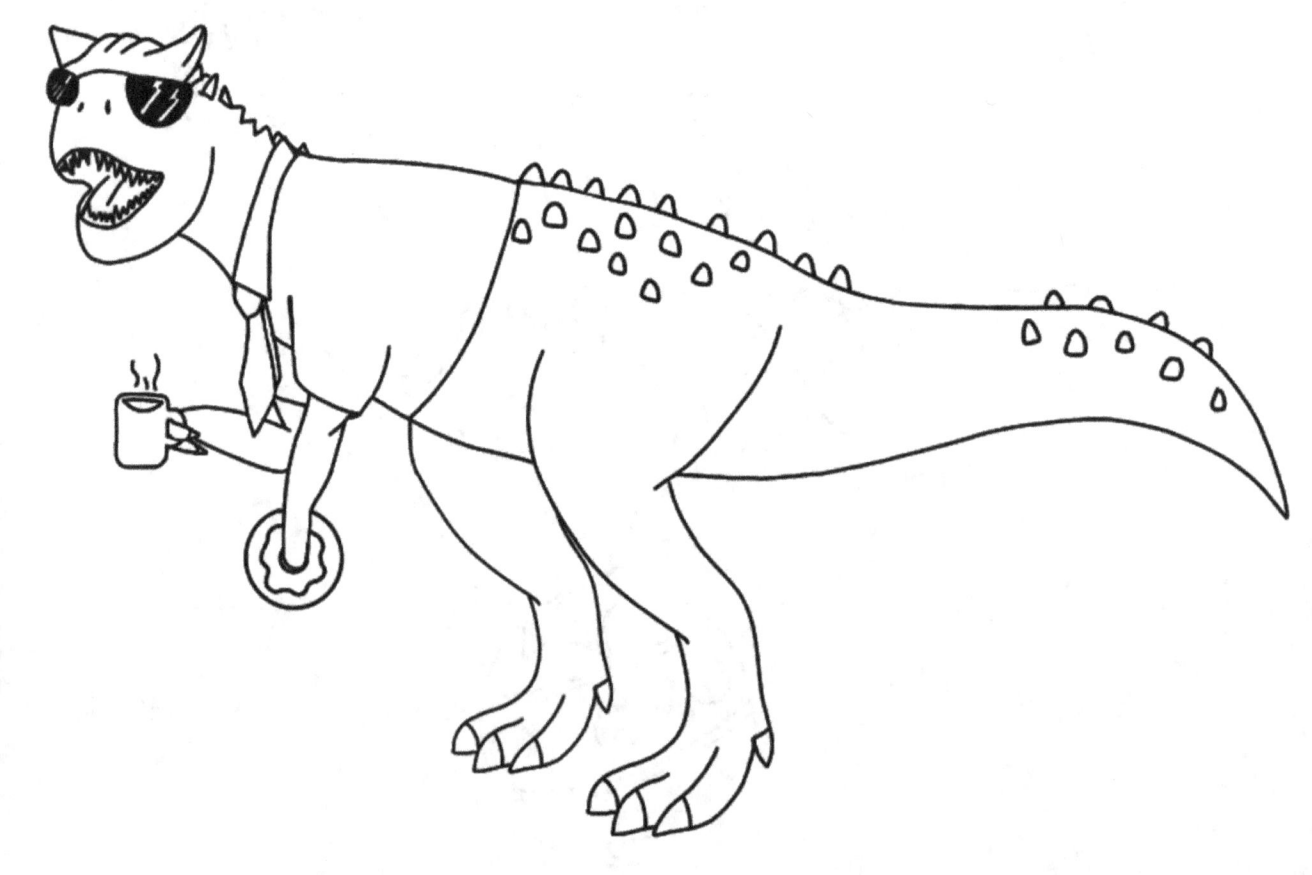

I AM
Creative

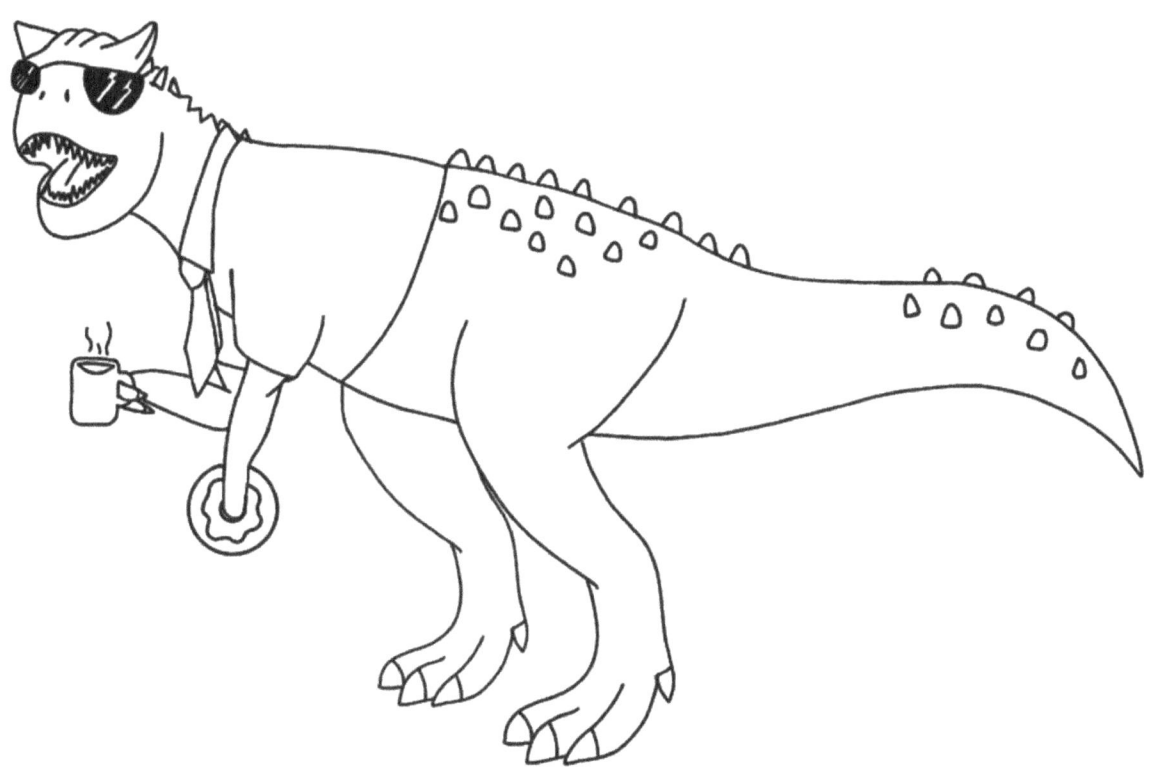

I AM
Confident

I AM
Helpful

I AM
Polite

I AM
Enough

I AM
Thankful

I AM
Patient

I AM
Happy

I AM Unique

I AM
Energetic

I AM
Brave

Dino Fun

Bonus Dino Activities!

Dino Counting

Pick a number card, count out the Dinos and put them in the garden.

Number cards and Dinos are on the next page.

Dino Counting

Cut out the number cards to place in the box

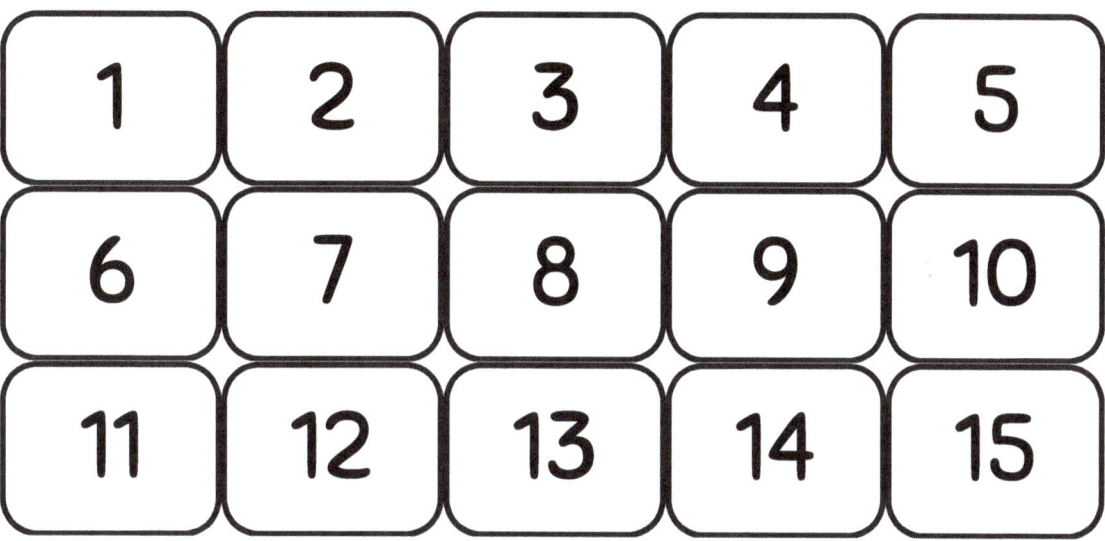

Cut out the Dino's to place in the garden.

Other Books By Dr. Joy Semien

Books in The Series
Learn to Count With Dino & Friends
Learn to Write With Dino & Friends
Learn to Read With Dino & Friends
A Dino & Friends Activity Book

Other Children's Books
A K.A.P.S. Guide to Preparing for Disasters: Children's Edition
The Giant and the Fairy

The End

www.ingramcontent.com/pod-product-compliance
Lightning Source LLC
Chambersburg PA
CBHW062226220526
45471CB00009B/3356